LAND SHAPES

D0211914

INDEX
GLOSSARY
ACTIVITIES
& PARKS

to volumes 1 to 13

Grolier Educational Corporation
SHERMAN TURNPIKE, DANBURY, CONNECTICUT 06816

Editorial
Brian Knapp, BSc, PhD
Art Director
Duncan McCrae, BSc
Production Controller
Gillian Gatehouse
Print consultants
Landmark Production Consultants Ltd
Printed and bound in Hong Kong
Produced by
EARTHSCAPE EDITIONS

First published in the USA in 1993 by
GROLIER EDUCATIONAL CORPORATION,
Sherman Turnpike, Danbury, CT 06816

Library of Congress #92–072045

Cataloging information may be obtained
directly from Grolier Educational Corporation

Title ISBN 0–7172–7189–7

Set ISBN   0–7172–7176–5

CONTENTS

Using this master index

These are the first 12 volumes
of the Landshapes set:

1: Mountain
2: River
3: Valley
4: Canyon
5: Beach
6: Dune
7: Geyser
8: Cave
9: Glacier
10: Volcano
11: Reef
12: Lake

Example index entry:

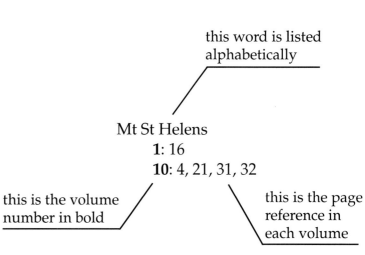

this word is listed
alphabetically

Mt St Helens
1: 16
10: 4, 21, 31, 32

this is the volume
number in bold

this is the page
reference in
each volume

Index

A

aa-aa lava
10: 17, 36
13: 20
abrasion
6: 15, 36
13: 20
acid
8: 13
Africa
10: 13
Agathla Peak
10: 28
algae
7: 19, 23
11: 4, 9, 14
Alaska
13: 32
Alps
3: 26
Amazon
2: 4, 34
Andes
1: 4
2: 34
10: 4
12: 24
Antarctica
9: 32
Anthias
11: 33
Appalachians
3: 2
Arabian desert
6: 3
Arizona
13: 32
Aral Sea
12: 3

arch
4: 8, 16, 23
Arches National Park
4: 2, 4, 16
arete
1: 29
9: 23
ash
1: 16
10: 8, 10, 12, 20, 21, 23, 30, 33, 36
13: 20
Aswan
2: 32
Atacama desert
6: 2
Athabasca Glacier
9: 34
Atlantic Ocean
10: 13
atoll
11: 10, 22, 24, 25, 28, 30
Australia
11: 4
12: 34
avalanche
3: 15
9: 24

B

backwash
5: 20, 22, 36
13: 20
bacteria
7: 30
Bali
12: 31
Banff National Park
1: 32
bar
2: 25, 36
13: 20
barchan
6: 24

barrier beach
5: 34, 35
barrier reef
11: 10, 22, 24, 25, 26, 27
barriers
11: 30
basalt
10: 18, 19
basalt column
10: 19
basin
3: 10
bats
8: 32, 34
bay
5: 10, 19, 34
beach
11: 20
12: 12, 20, 21
Bermuda
11: 2
Big Room, Carlsbad Caverns
8: 33
billabong
12: 33
Black Sea
12: 3, 34
blizzard
9: 24
block mountain
1: 12, 14, 36
13: 20
Blue Nile
2: 32
Bodensee
12: 3
bomb
10: 24
Bora Bora
11: 3
Borneo
8: 3
boulder
2: 14

box canyon
    4: 30, 34, 36
    13: 20
Brahmaputra River
    3: 4
braid
    2: 22, 36
    13: 20
breaker
    5: 19, 20, 25
Bumpus Hell
    7: 28
butte
    4: 11, 22, 25, 26, 36
    13: 20
butterfly fish
    11: 14

C

Cairo
    2: 33
California
    13: 33
Canada
    1: 32
    9: 34
canyon
    3: 8, 31
    6: 12, 36
    13: 20
Canyonlands National Park
    4: 30
Cape Hatteras
    5: 35
carbon dioxide
    7: 29
    8: 13, 24, 36
    13: 20
carbonic acid
    8: 13, 36
    13: 20
Caribbean Sea
    11: 2

Carlsbad Caverns
    National Park
    8: 2, 32
Cascade Range
    10: 31
Caspian Sea
    12: 3, 4, 8, 34
Castle Geyser
    7: 19, 21
cataracts
    2: 32
cave
    4: 17
    8: 8, 20, 36
    9: 10, 18, 19
    11: 35
    13: 21
cave decorations
    8: 21
cave entrance
    8: 16
cavern
    8: 20
cay
    11: 20, 22
cement
    4: 17, 29, 36
    6: 13, 36
    13: 21
Cerro Aconcagua
    10: 4
chamber
    8: 8, 20
    10: 14, 21
channel
    2: 9, 10, 36
    3: 28
    4: 24
    11: 20
    13: 21
chaos
    6: 18, 36
    13: 21

Charles Darwin
    11: 24
Chimborazo
    1: 4
cinder
    10: 24, 25
cirque
    1: 27, 36
    9: 11, 22, 25, 36
    12: 23, 36
    13: 21
clay
    2: 28, 30
cliff
    4: 8, 20, 27
    5: 11, 16
    12: 12, 20, 21
clint
    8: 12, 36
    13: 21
Clipperton Island
    11: 25
Colorado
    4: 4, 30, 35
    13: 33
colors
    7: 19
Columbia Icefield
    9: 34
Columbia River
    10: 19
Columbia River Gorge
    10: 19
column
    8: 8, 10, 28
cone
    7: 20, 36
    10: 10, 13, 24, 25
    13: 21
continent
    10: 13
continental shelf
    11: 11, 36
    13: 21

rock pillar
4: 22, 28, 29
rock step
9: 16, 17, 18
Rocky Mountains
1: 32
7: 14

S

Sahara desert
6: 3, 4, 27, 32
salt lake
12: 28, 34
San Agustin
8: 2
San Juan River
4: 12
sand
2: 14, 15, 17, 28
5: 9, 14, 17, 24, 36
6: 8
12: 8, 29
13: 28
sand-blasted
6: 15, 36
13: 28
sand buggies
6: 23
sand dune
5: 11, 28
sand grains
6: 13, 15
sand mountains
6: 11, 28, 34
sand sea
6: 8, 30, 36
13: 28
sand spit
5: 11, 36
13: 28
sandstone
1: 10
4: 19, 20, 24, 32

sandstorm
6: 14, 15, 36
13: 28
sandy beach
11: 20
Santa Cruz
11: 25
Saskatchewan Glacier
9: 34
scale
7: 17, 20
scar
3: 15
Scotland
3: 3
12: 24
scour
3: 23
9: 26
scratch
9: 21
scree
1: 24, 36
3: 14
13: 28
Sea of Gallilee
12: 34
sea stack
4: 27
Sechura desert
6: 2
sediment
1: 10
12: 13, 16, 36
13: 28
seif dunes
6: 26, 27
serac
9: 17, 36
13: 28
shaft
8: 17, 19
shale
4: 19, 20, 24, 32

shelf atoll
11: 29
shelf reef
11: 22, 30, 32, 36
13: 28
shell
5: 12
shoals
2: 15
sill
10: 14, 36
13: 29
silt
2: 14, 17, 28
11: 21, 26
Simpson desert
6: 3
sink
8: 10
sinter
7: 21, 36
13: 29
slag
10: 15
slate
1: 10
slip
3: 15
slope
4: 20
snorkelling
11: 17, 27
snout
9: 20, 29, 36
13: 29
snow
9: 10, 12
snowball
9: 12
snowflake
9: 9
snowmelt
7: 30
soda straw
8: 24

Glossary

aa-aa lava
this is a kind of lava that is quite sticky and which, as it cools, moves with difficulty. Its surface tends to break up into piles of broken rock

abrasion
the wearing away, or scouring of a rock as sand grains are thrown against it during a sand storm

ash
the very fine material that falls from clouds above a volcano. Much of the ash would have started out as fine spatters of liquid lava which cooled in the air. The rest would be fragments of shattered rock

backwash
the rush of water that flows down a beach after the wave has reached its limit of advance. Much of the surf sinks into the beach, so the backwash is only a small part of the water that rushes seaward

bar
a deposit of pebbles or sand that builds up in slack water on the inside bank of a curve in a river channel

block mountain
a mountain formed when a large block of crust is pushed up or when surrounding blocks move down. The block mountain is surrounded by shear sides made by repeated faulting

box canyon
the name given to a small canyon that is tributary to a large canyon. Box canyons end in steep-sided cliffs

braid
a temporary island that splits up the water in a river channel. Braids occur in large numbers and are changed in size and shape by each flood. By contrast, islands occur in small numbers and do not change after floods

butte
a natural tower of considerable size and which is no longer part of the main tableland. There is no special size for a butte. It changes to a pillar rock when the butte is nearly worn away

canyon
a gorge-like valley found in dry areas. Canyons have bare, rocky sides with no soil cover. Other names for canyon include arroyo (US) and wadi (north Africa)

carbon dioxide
a gas which is found in the atmosphere. Carbon dioxide is also produced by animals living inside the soil. Rainwater absorbs carbon dioxide, forming a weak acid which can then dissolve limestone rock

carbonic acid
a chemical which is able to dissolve limestone and widen cracks between limestone blocks into passageways and caverns. Carbonic acid is formed as carbon dioxide gas is absorbed by rainwater

cave
a large underground network of passages and chambers formed by natural processes and which are big enough to be explored. Most caves are dissolved out of solid limestone by the slow action of water. People sometimes use the word 'cave' as a shorthand for 'cave system'

cement
the natural material that holds sand grains and other particles together to make a rock. Lime is a common cement and is easily dissolved when it gets wet, allowing the sand grains to be washed or blown away

Many rocks are stuck together with natural cements as they are being formed deep under the sea. One of the most common cements is natural lime. When a rock with a weak cement is exposed to the weather, the natural lime cement dissolves and the sand grains fall away. This is an important reason for the growth of arches

channel
the shallow trench that has been cut by a river or stream

chaotic
from chaos, a word meaning something that has no pattern or sequence to it

cirque
a bowl-shaped hollow that is scoured out of a mountainside by a glacier. Other common names include corrie and cwm. Cirques begin as stream-cut hollows in mountainsides. In the Ice Age these areas filled with snow and ice. As the ice flowed downhill it cut fastest in the center of the hollow, where it was deepest. After the Ice Age ended many hollows filled with water to make lakes. Many countries have a separate name for the lakes formed in the hollows. One such word for a mountain lake is tarn

clints and grykes
A clint is a block of limestone and a gryke is the widened crack between blocks

cone
the dome-shaped build-up of sinter that forms at the place where some geysers erupt

continental shelf
the edge of a continent that happens to be below sea-level. Most continental shelves are broad and slope only gently. Continental shelves in the tropics are common along the Red Sea, East African, East Asian, Caribbean, Brazilian and Australian coasts. They are rare along the west coast of America because this coast has many deep ocean trenches

coral
the coral that makes sand in many tropical places comes from broken pieces of coral reef. Coral reefs are made from small animals called corals that make delicately branched skeletons

corrode
the acid action of gases and hot water that can dissolve rocks and wear away passages for the water to move through. Most geysers, hot springs, mud pots and fumaroles make their passages wider with this hot acid

crater
  (i) the depression that forms in the center of a volcano. Sometimes craters fill with water to give crystal-clear lakes;
  (ii) the pit-shaped depression that forms at the place where many fountain geysers erupt

crevasse
a wedge-shaped chasm in the surface of a glacier. Crevasses show that the ice is brittle and that it will crack near the surface, but lower down the pressure is greater and the ice flows, so crevasses cannot form

crust
the name for all the rocks that make up the solid surface of the Earth, usually the top 10 to 100 mi. The rocks of the crust are hard, cold and brittle and will crack when pulled or pushed. The rocks may seem to be solid and stationary to us, but over a long period of time the rocks of the crust move up and down and also across the surface of the Earth. The forces that make them do this are in the part of the Earth below the crust. This region, called the mantle, is molten and turns over and over in much the same way as soup moves when it is heated in a pan. The crust is broken into a number of very large slabs, known as plates. The continents are the part of the plates that show above sea level

delta
the name for a fan-shaped wedge of sediment that builds up in a lake or the sea as a river drops its load of silt in still water. Deltas are similar to icebergs in so far as only a small part of their true size stands above water. Deltas give a small idea of just how much material is brought down by the world's largest rivers

deposits
any material that has been carried by a river and which is then dropped. Clay, silt, sand and pebbles are all materials dropped, or deposited, by a river at times of slack water

dissolve
a material which can be absorbed by a liquid without showing any visible signs of its presence. Rainwater contains many invisible substances that have been dissolved in its passage through air, soil and rock

divide
the line that separates one valley from its neighbor. It is found by looking for the highest points on the ridges of hills between valleys

dormant
a volcano which has not been active for many years, but which scientists think will erupt in the future. Some volcanos may only erupt once every thousand years. It is quite difficult to know whether a volcano is dormant or extinct

dune

a dune is a mound of sand that has been dropped by the wind. There are many shapes and size of dunes, mostly found in deserts. Coastal dunes are often colonized by grasses

dust

the fine pieces of rock waste that are easily caught by the wind. People sometimes use the words clay and silt to mean dust. When dust becomes wet it turns into mud

dyke

a sheet of lava that once cut across layers of rocks and then turned into a solid. Dykes are found as ridges in the landscapes in many countries. They show that volcanos were once active in the area

earthquake

the name given to violent ground shaking. It is caused by rocks moving past each other along a fault. The strength of earthquakes is measured on a scale called the Richter scale

ecosystem

the pattern of living creatures that exist side by side in an area and which depend on each other. Often life in an ecosystem is made into a food chain, with plants giving food to grazing animals, which in turn are food for hunting animals. When the animals die their remains rot and become food for the plants again

energy

the ability to do work. The way that, for example, the energy of the wind can be transferred to waves and therefore drive water onshore. Energy is always transferred, never used up. So when waves break on a beach and the water comes to rest, the energy is transferred to the sand, which then moves. In a river energy comes from the amount of water flowing and the speed of the water. The rivers with the largest energy - and those which can cut swiftly into the landscape - are large and fast-flowing

erode

the wearing away of the land by rivers, waves, wind, landslides and ice. There are two parts to erosion: first the rock must be loosened, and then it must be carried away. A river erodes its banks because it loosens the material and carries it away instantly. However on a slope frost may loosen a rock but it may be many years before it falls away

(i) Rivers mainly erode their beds by using the sandpaper-like action caused as pebbles and sand bounce along the bed. The force of the water alone is enough to erode the bank of a river

(ii) Erosion of a cliff takes place in two stages. First the rocks of the cliff are weakened by the constant pounding of waves, then they are carried away by the waves. Beach sand is eroded in one stage because it is already loose material

(iii) Glaciers erode by scraping and plucking as they slip over the land

erosion cycle

the sequence of events that occurs when rivers, landslides, rock falls and other processes erode an area of high land, gradually reducing it to a plain

erupt
the pattern of events that happen when a volcano becomes active. First it causes earthquakes, then the vent is blown clear and finally ash and/or lava come out of the vent. Eruptions usually only last between a few days and a few months

estuary
the mouth of a river as it enters the sea. Estuaries occur in low-lying areas and they are usually filled with sand and mud

etch
a word used to describe erosion which picks out the tougher and weaker rocks clearly

extinct
the name given to a volcano that scientists think has stopped erupting forever

face
the clear-cut edge of a rock

fault
a break in the rocks of the crust. Faults sometimes cause land to be lifted into mountain ranges

fissure
the name for a deep crack in the crust that may go all the way to the magma deep below the Earth's surface. When lava cools in a fissure it makes a dyke; in a geyser, fissures allow heated water to get through the rocks and corrode them

flash flood
a sudden storm in a desert may cause large amounts of water to flow from rocky hills and gather on the nearby plains and in river beds. There is no time for the water to sink in to the ground, and instead it floods out across the countryside in a matter of minutes, often without any kind of warning

flood plain
the area in the bottom of a valley that is covered with water during a flood. Flood plains are made of material deposited by rivers

flowstone
(i) the general name for all the deposits other than stalactites and stalagmites that have been formed as limestone comes out of solution. Most flowstone makes a kind of drapery on cave walls and ceilings
(ii) the name given to the deposits of limestone from hot springs and pools which builds up in smooth flowing shapes to cover the surface rocks below some hot springs. It can look like a frozen waterfall

fold mountain
a mountain which has been formed as two of the Earth's plates move together, crumpling up the rocks in between

fumarole
the name for the rising steam that comes from some openings in a geyser field. Fumaroles only spout steam because there is not enough water for pools or geysers

glacier
a river of ice that flows from an ice-filled hollow or an ice cap high in mountains. A much larger body of ice would be called an icecap or an ice sheet

grains
particles of rock waste, often of a single type of material, that are small enough to be readily moved by wind or water. Sand particles are called grains; bigger particles such as gravel or pebbles, are not considered to be grains

ground moraine
the debris carried underneath a glacier. It has been plucked from the valley floor

Ice Age
the time, from a million years ago, when the world's climate became colder and ice sheets formed on lowland beyond the mountains on all the northern continents. At the maximum of the Ice Age ice sheets covered nearly a third of all the land in the world. The ice sheets only melted away a few thousand years ago, which is the reason so many scoured landscapes still show the effects of glaciers so clearly

(i) Lake hollows were made in the early stages of the Ice Age, but they were also made near the end when glaciers could no longer carry their debris and they dropped it in irregular ridges across the land. Ice-formed lakes are common because the ice sheets and glaciers melted away only a few thousand years ago

(ii) As more and more ice formed on land, so rivers no longer returned rainfall to the seas, and ocean levels began to fall. So much water was locked up as ice that the sea levels throughout the world fell by 240 ft and glaciers were able to cut valleys into what is now the sea bed. When the Ice Age ended the ice melted and the sea level rose again, flooding these ice-cut valleys and making, for example, fjords

(iii) During this time ice sheets swept soil from many areas and left limestone rocks bare. Since the ice retreated a few thousand years ago there has not been time for new soil to form

ice sheet
a large body of ice, often covering lowland but also completely burying mountains, which occurs only in the world's coldest places such as Antarctica and Greenland

icefield
a large body of ice, usually high up in the mountains. Icefields supply the ice for many valley glaciers

labyrinth
a maze of twisting and turning underground passageways

landslide
a rapid movement of a large slab of soil and rock on a valley side

lateral moraine
the name for debris that falls from mountainsides and comes to rest on the edges of a glacier

lava
the molten material that flows from a volcano when it is erupting. Lava is orange or red in color when it is hot and molten, but it soon cools to a black color. The rock is often full of trapped bubbles

levee
a ridge of sand and silt that builds up on the edges of a river channel. It is formed during floods when the river spills over its banks. As the fast flowing water spreads over the nearby land it slows down and drops its load, forming a levee. The Mississippi is famous for its levees which are several yards high

limestone
the rock that is formed entirely of the limey remains of sea creatures and which are made of a substance called calcium carbonate. Chalk is also calcium carbonate

magma
molten material that flows from deep within the Earth. It is the source of all materials that come from a volcano during an eruption

meander
a natural curve or bend that a river makes as water flows in a channel. River meanders change size and shape all the time as rivers cut into the outside curve and drop material on the inside. If the curves swing almost completely back on themselves, the meander is called a gooseneck

medial moraine
a ridge of surface moraine that has been made by the merging of two lateral moraines

mesa
the edge of a tableland that has been eaten into by many canyons

mineral
a natural material from a rock that can form crystals. The minerals dissolved in the water that comes from geysers and hot springs builds up deposits. Tiny crystals can be seen in these deposits

moraine
the name for any material that is carried by a glacier or ice sheet. Most of the material is carried along the bottom of the glacier or ice sheet and is dropped over the land when the ice melts

mountain chain
a series of mountain ranges stretching thousands of miles across a continent

mountain range
a single narrow ridge of mountains separated from other ranges by valleys

mouth of river
the place where the river reaches the sea. Sometimes the river mouth is deep and wide, this is called an estuary. Other rivers build up deltas at their mouths

mudflow
a rapid movement of sodden soil or soft clay rock on a valley side

nozzle
the name for the build-up of minerals around the place where the geyser spurts from the ground. Geyser shapes affect the way the geyser erupts

oxbow lake
a small curved lake that shows where a meander used to be. Oxbows are formed

when these meander curves become so pronounced that they cut a complete meander loop off

pahoehoe lava
the runny kind of lava that flows for quite long distances before it cools and hardens. It often forms rope-like patterns on its surface

pavement
a name given to large areas of exposed flat limestone rock where the widened cracks and blocks can easily be seen. The largest area of limestone pavement in the world is in western Ireland, although this is not a region of large caves

plain
a region of low lying land that has been produced by river action. Plains are not absolutely flat, but have many gentle slopes on their surfaces

plankton
the smallest forms of plant and animal in the sea. They are carried along by the ocean currents, using minerals in the sea water to build their cells. They are an important food source polyp the soft-bodied anemone-like organism of a coral. Polyps start life as tiny free-swimming larvae which find a suitable vacant spot to attach themselves, then produce a limestone skeleton. After a while they divide, creating two polyps where once there was just one. Within a short time further dividing has produced a large group, or colony, of coral polyps, each making their own skeletons

plate (continental)
large areas of the Earth's crust. The crust is made up of a number of plates which are constantly moving

plateau
a name for a tableland, or flat high land

platform
the broad, gently-sloping surface that surrounds a geyser. It is built from minerals that are deposited as the geyser water cools when it flows away from the nozzle

porous
a name used to describe a rock that has many small gaps between its rock particles and which can therefore soak up water

pothole
(i) the deeply scoured pits that are made as pebbles are swirled round and round on a river bed
(ii) a name for an entrance to a vertical cave shaft

profile of valley
a cross-section of a valley. This is the view you would get while looking upstream. Profiles simply show the shape of the land from the divides to the river

rapids
stretches where the river rushes and tumbles over exposed rocks

reef
a large bank of limestone produced by the skeletons of corals and other animals in a tropical sea

rift valley
a valley with very straight, steep sides that has been made as a region of the Earth's crust has been pulled apart, allowing blocks of crust to drop down. Rift valleys have steep sided straight valleys and flat floors

river cliff
the steep bank on the outside curve of a meander. It is made steep by the scouring action of the water

sand
sand is the name for a special size of particle which can be made of many kinds of materials. Mineral sand is mainly made of small pieces of quartz rock that have been broken down. Other types of sand include coral and shell sand, both the broken remains of animal skeletons

sand-blasted
the process of cleaning metal in a factory using sand grains blown from a high pressure air gun. Nature can produce much the same effect during a sand storm

sand sea
the large areas of sand dunes that are found in the world's great deserts. Sand seas would normally cover many thousands of square miles

sand spit
a long ribbon of sand that stands clear of the coast, but is linked to it at one end. Sand spits usually form when there is a marked change in the direction of the coastline

sandstorm
a time when the winds blow so strongly that sand grains can be carried along. A duststorm is far more common than a sandstorm, and dust is always lifted into the air when winds are strong enough to move sand. Dust gets in the eyes and mouth, but it does not erode the land; sand scours rocks up to about six feet above the general level of the plains

scree
broken rock that has fallen from a mountain and which builds up into a fan-shape in the valley below

sediment
the name for any material that has been carried in suspension in water and later dropped on the bed of a river or lake. Mud is the smallest size of sediment, silt is a little larger, with sand and finally pebbles as the largest materials

serac
a knife-edged ridge that forms between two crevasses in places where the ice surface is very disturbed

shelf reef
a reef that forms on the broad, shallow waters of a continental shelf. Many small coral islands may rise from the reef and be dotted over the sea. This pattern is different from island reefs, which usually form in long lines across an ocean

sill
a sheet of lava that forced its way in between other layers of rock

sinter
the name given to the quartz scale that makes nozzles and platforms. It is made from the mineral quartz, the same mineral found in some types of sand

snout
the end of a glacier or ice sheet and usually the place where the ice melts away

soil creep
the imperceptibly slow movement of soil down a slope. Soil creep is the most important soil-moving process on soil-covered slopes

solution
the process whereby solid materials are gathered and dissolved in a liquid. Calcium carbonate is brought into solution when it reacts with weak carbonic acid

source
the place where the river first begins to flow. Some rivers begin as flowing springs. More commonly rivers start with water seeping through soils to form a muddy patch

splay
the spreading out shape that occurs near the end of a valley glacier especially if the ice pushes out on to lowland where there is no valley to hold it in a narrow tongue

spouter
a geyser which continuously boils over without ever really making a tall plume of water. It is half way between being a pool and a true geyser

spring
a place where water seeps from a rock. If a spring occurs high up on a cliff, the cliff is said to weep

stalactite
a long cone-shaped formation that is found hanging from the roof of a limestone cave. It is produced as lime-rich water slowly drips from the roof, leaving behind the lime as a kind of scale

stalagmite
a long cone-shaped formation that grows upwards from the floor of a cave. It forms below a point where water drips from the cave roof. Stalagmites are much broader and more massive than most stalactites

striation
a scrape-mark made on the surface of rock where glaciers have recently been

superheated water
water above its normal boiling temperature. The water seeping through the rocks 2000 m below the surface is under so much pressure that it may get as hot as 400 °F and still be a liquid. (At the surface water boils and becomes steam at 212 °F). When superheated water reaches the surface it almost explodes into steam. This is what gives the geyser its power to jet high into the air

surface tension
the force that exists when water partly fills a grainy material like sand. The force pulls the grains together, but for it to work the sand must be damp, not completely wet

swamp
a region where non-moving shallow water is found and which has been colonised by plants so that, from the air, it looks like land. Swamp is often the name given to wetlands in warm areas and in cooler regions they are commonly referred to as a marsh or bog

swash
the name given to the part of the breaking wave that rushes up the beach. The foaming swash is very powerful and can easily damage reefs, but it also carries broken reef material across towards inner lagoons

tableland
a name for a large area of high level land. Another name for such a land is a plateau

trench
a gorge cut into the land by river action

tributary
(i) small streams or valleys that cut into a landscape and help deliver surplus water to the main river or valley
(ii) a river or stream that feeds into a bigger river

U-shaped valley
the deep trench cut by a glacier as it flows from a mountain

vent
the central pipe of a volcano leading from the magma chamber to the surface

volcano
a place where lava and ash are erupting at the Earth's surface. Lava and ash usually build up around the eruption, making a cone-shaped mountain

waterfall
a place where a river spills over a level band of rock and drops through the air. The smallest waterfalls are no more than steps a yard or so high, the largest are nearly 3000 feet high

weathered rock
a rock is weathered by elements such as frost, sun or rain, causing it to break up or fall apart

Activities

Volume 1, Mountain

Volume 2, River

Volume 3, Valley

Volume 4, Canyon

Volume 5, Beach

Volume 6, Dune

Volume 7, Geyser

Volume 8, Cave

Volume 9, Glacier

Volume 10, Volcano

Volume 11, Reef

Volume 12, Lake

Where to find areas of outstanding landscape beauty in the US

This listing contains some of the National Parks, National Monuments, National Recreation Areas and other protected sites that you can visit when on vacation. Most landscapes contain more than the features listed here, so always look out to see what else you can find.

NL = National Lakeshore

NM = National Monument

NP = National Park

NRA = National Recreation Area

NS = National Seashore

ALASKA

Aniakchak NM
A region of active volcanos and craters

Glacier Bay NM
As glaciers reach the coast they calve and produce icebergs

Katmai NM
The area which erupted in 1912 to leave a region of smoking ash called the Valley of Ten Thousand Smokes

Kenai Fjords NM
A coastal area with mountains, glaciers and fjords

Kobuk Valley NM
Sand dunes on the move

Lake Clark NM
Mountain ranges with volcanoes, mountain peaks, glaciers, valleys, lakes and waterfalls

Mt McKinley NP/ Denali NM
A huge wilderness area of glaciers, mountains, valleys and rivers. Access is difficult

ARIZONA

Grand Canyon NP
Hike down into the world's largest land chasm

Sunset Crater NM
Volcanic cone

CALIFORNIA

Golden Gate NRA
The seashore area next to San Francisco and containing spectacular cliffs by the Golden Gate and semi-circular Half Moon Bay farther south

Death Valley NM
A huge rift valley sunk between high rugged mountains. Sand dunes, salt lakes, volcanic craters are all to be found in this hot desert. Best visited in the winter!

Devils Postpile NM
A giant column of basalt rock

King Canyon NP/ Sequoia NP
Rugged mountains, valleys and rivers. These are mainly hiker's parks

Lava Beds NM
Spectacular sheets of lava that have been produced by vast outpourings and left a broken and wild land of deep ravines

Mt Lassen Volcanic NP
This volcano erupted at the beginning of the century. Ash and cinder cones can still be seen, together with many hot springs and mud pots

Point Reyes NS
Dramatic wave-washed beaches. Here you can also see the effects of the San Andreas (earthquake) Fault on which nearby San Francisco is built

Yosemite NP
One of the worlds greatest natural wonders and most visited of the NPs. It contains glaciers, waterfalls, ice-scoured mountains such as famous Half Dome, rivers, deep valleys, El Capitan cliff and the remains of a lake.

COLORADO

Black Canyon of the Gunnison NM
Deep canyon cut into the mountains

Great Sand Dunes NM
Sand seas including the highest dunes in the US

Mesa Verde NP
Spectacular cliff landscape (also famous for its Indian dwellings)

Rocky Mountains NP
Part of the Front Range of the Rocky mountains shows rugged peaks and cirques, rivers and lakes

FLORIDA

Everglades NP
This is one of the best preserved deltas and swamplands on the US

HAWAII

Haleakala NP
The volcanic crater on Maui

Hawaii Volcanos NP
Kilauea and Mauna Loa are both active volcanos and add an extra sense of excitement to this volcanic mountain park

IDAHO

Craters of the Moon NM
See sheets of black lava intermingle with cindery craters. There are also some spectacular lava caves

INDIANA

Indiana Dunes NL
On the edge of Lake Michigan this area has giant dunes and sandy beaches

KENTUCKY

Mammoth Cave NP
The world's longest known cave system can be visited for tunnels and cave formations

MAINE

Acadia NP
Rocky seashore of the Schoodic Peninsula

MASSACHUSETTS

Cape Cod NS
Beaches along the outer cape, an area formed from glacial moraine

MINNESOTA

Voyageur NP
An area of lakes in the heart of the country

MISSOURI

Ozark National Scenic Riverways
See the winding lowland shapes of the Current and Jacks Fork rivers

MONTANA

Glacier NP
Centered around an amazing valley that was cut through the mountains by glaciers. Mountains, glaciers, waterfalls, rivers and lakes

NEVADA

Lehman Caves NP
One of the world's best collections of caves with stalactites and stalagmites. Above the caves there are high wild mountains

NEW MEXICO

Capulin Mountain NM
The crater of a volcano

Carlsbad Caverns NP
Huge caverns and wonderful cave formations. An elevator brings you to the surface after miles of walking underground. (Also famous for its bats, seen flying out at sunset)

White Sands NM
A dried up lake bed with dunes made of gypsum

NORTH CAROLINA

Cape Hatteras NS
Beaches and offshore barrier islands at their best on the Atlantic coast

Cape Lookout NS
A remote area of offshore barrier islands

OREGON

Crater Lake NP
A circular lake surmounting a huge volcano whose top was blown away thousands of years ago. Lake, island, cliff, volcano

Oregon Caves NM
A system of caves showing many tunnels and other water-carved passages

SOUTH DAKOTA

Badlands NP
Semi-arid land scoured by rivers to give a uniquely barren dissected landscape

TENNESSEE

Great Smoky Mountains NP
The rounded summits of the high Appalachians form the backbone of this area with valleys and rivers. (This is the most visited of the National Parks)

TEXAS

Big Bend NP/ Rio Grande Wild and Scenic River
The area around the Rio Grande has mountains, gorges and sweeping meanders

Guadaloupe Mountains NP
An area of limestone and an ancient reef now high and dry

Padre Island NS
One of the hundreds of barrier islands that hug the south and east coasts of the US. This one is a bird sanctuary

UTAH

Timpanogos Cave NM
See the underground world of caves

Arches NP
Water-worn sandstone cliffs are pierced with intriguing arches of all shapes and sizes

Bryce Canyon NP
Soft rocks have been sculpted into tall pinnacles in a cliff edge

Canyonlands NP
The place to see the great sweeping bends of rivers as they cut into tablelands

Capitol Reef NP
Huge ranges of mountains have been bucked by Earth movements. In this semi-arid area there is little vegetation to mask the features

Zion NP
Tall sandstone cliffs trap a river in a deep chasm

Rainbow Bridge NM/ Natural Bridges NM
Spans of sandstone arch over a river in spectacular fashion

VIRGINIA

Shenandoah NP
a broad ridge and valley country that shows many features of valleys and rivers in the more humid east of the US

WASHINGTON

Mt Ranier NP
Huge volcanic mountain surmounted by an ice cap. Moraines, ice falls, glaciers, rivers, deep valleys

Mt St Helens NM
The famous volcano which erupted in 1980. The effects of the blast can still clearly be seen

North Cascades NP
High ice-scoured mountains with glaciers, icefalls, cirques, mountains, rivers, waterfalls Ross Lake NRA and Lake Chelan NRA are linked to the NP and contain many similar features

Olympic NP
A spectacular coastline and mountain range with cliffs, beaches, waves, mountains, rivers (an area of very heavy rainfall)

WYOMING

Yellowstone NP
This is the place to see the world's greatest geysers, but it also has rivers, lakes, waterfalls and many volcanic features

Devils Tower NM
A huge tower made from the vent of a volcano

Grand Teton NP
The Rocky Mountain front shows at its most stupendous. Mountains, glaciers and lakes